Samsung Galaxy s21 5G Series user Manual

I0481256

Knowing your device

The comprehensive user manual to help you steer through

Ella Brown

Contents

Introduction

Ultra - where you definitely end up when we one-up our usual Plus. We are saying, Samsung's bested itself this year and has gone beyond its S21+ to give us the Galaxy s21 5G series.

While leading the roster this year, the Ultra made a shift when it is compared to the 2020 lineup. Whereas we had the similar the previous year in many ways S10 and S10+, the S10e positioned beneath them, but now there's a model which ranked above the typical S21 and S21+ - the new autograph makes all the sense then.

The s21 5G series has an even bigger display and packs, a larger battery with faster charging mode than the others, plus it even comes with up to a tremendous 16GB of RAM.

The camera is probably the most extraordinarily built we've seen to date. The main cam uses a ample 108MP sensor that combines 9 small pixels into a big one with dispensational designed to use the extra information being gathered - Samsung calls it Nonacell

the telephoto shooter – this has an unmatched blend of a large 48MP imager with a periscope lens presenting 4x optical zoom above the main..

There is another ultra camera touch over the front - Samsung's fitted a 40MP selfie unit to to make it unique from the regular 10MP units on the ordinary S21s.

Read along as you get to know your device.

SAMSUNG S21 5G

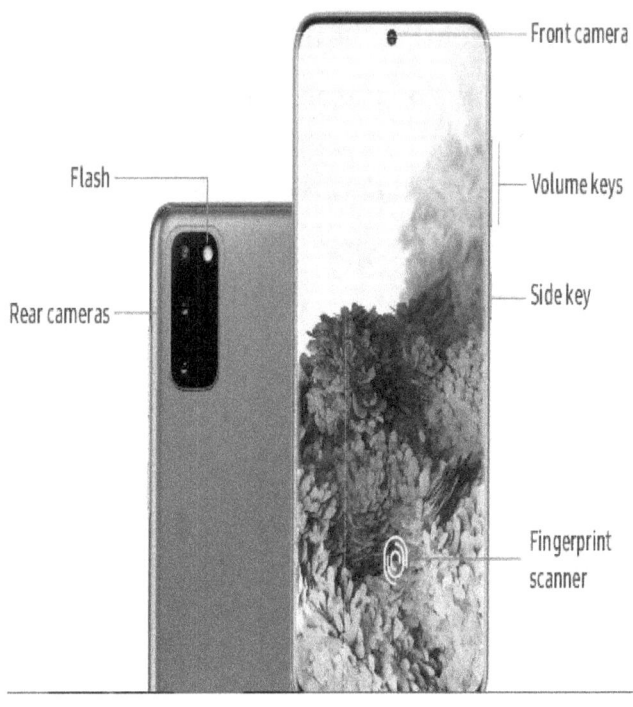

Front camera

Flash

Volume keys

Side key

Rear cameras

Fingerprint scanner

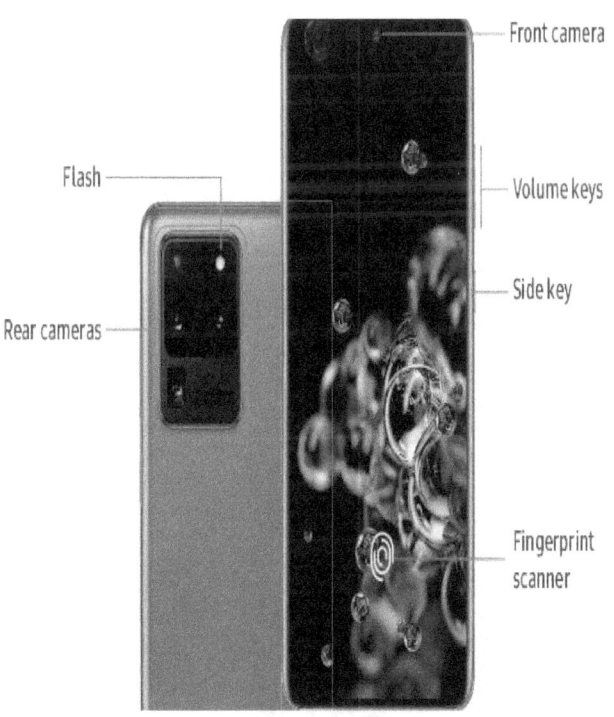

Front camera

Flash

Volume keys

Side key

Rear cameras

Fingerprint
scanner

Set Up your Device

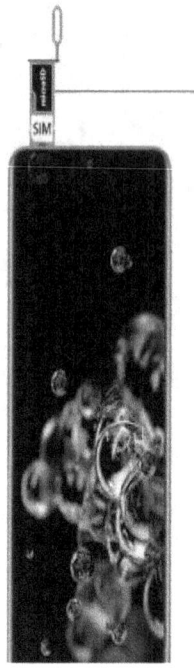

Install SIM/microSD card

Place the SIM card and optional microSD card (sold separately) into the tray with the gold contacts facing down.

A Nano-SIM card is usually used by this device. You may choose to preinstall a SIM card or even use your previous SIM card. However, your carrier's specifications and network availability determines the network indicators for 5G service. For more details, simply contact your carrier.

Note: Ensure it is only batteries and other charging devices that are approved by Samsung you use. If you want to maximize battery life, then you should use Samsung charging devices and batteries. If you are using other charging devices and batteries, then you may cause damage and void your warranty.

Note: The Samsung S21 is IP68 rated for water and dust resistance. Ensure that the Memory card and SIM card tray openings are kept or preserved free of water and dust if you want to maintain the dust-resistant and water-resistant features. Also, ensure the tray is securely and safely inserted prior to liquids.

Charge the Battery

Normally, a rechargeable battery powers your device. Besides, a charging head and USB Type-C cable are needed before you can charge your device from a power outlet.

TIP: If you experience that the device and charger becomes hot and eventually stop charging, simply know that it has no effect on the device's lifespan or even performance. What you should do rather is to disconnect the charger from the device and allow the charger to cool down for some minutes. Visit Samsung.com/us/support/answer/ANS00076952 for more information.

Wireless PowerShare

With your phone, you can wirelessly charge your compatible Samsung devices. While sharing power, you may experience that some features are unavailable.

- Click Wireless PowerShare from quick settings to enable this feature
- Place the compatible device on the back of the phone while it is facing down to charge it. When charging starts, a notification vibration or even sound occurs.

Note: Most Qi-certified devices works with wireless PowerShare. Also, your device battery should be minimum of 30 percent before you can share. The device also determines the power efficiency and speed of charge. Besides, this feature may not work with other manufacturer's devices and some accessories. Finally, remove any cover from each device if you have issues connecting the device or even if charging is slow.

Network environment

If you want to experience the best results as you use the wireless Powershare, simply get rid of any cover or accessories before you use the feature. Wireless PowerShare may not work properly if you use any type of cover or accessory. Since the location of the wireless charging coil varies, you may have to adjust where it is placed to make a connection. A notification or even a vibration occurs when you start charging. The notification only indicates to you that there is a connection between the device and the power outlet. Depending on your network environment, call data services or reception if affected.

Customize your home screen

The starting point for navigating your S21 device is the home screen. In here, you can place your preferred widgets or apps and also set up more Home screens, choose the main home screen, remove screens, and even change the order of screens.

App icons

To launch an app especially from the Home screen, simply use app icons.

- Press and hold an app icon from Apps
- After that, remove an icon by clicking Add to Home
- Press and hold an app icon from the home screen
- Finally, click Remove from Home.

Note: You are not deleting the specific app if you are removing an icon. You are only removing the icon from the Home screen.

Use Folders

On Apps screen or even Home screen, simply organize app shortcuts in folders. See create and use folders for more information.

Wallpaper

Once you select a favorite video, picture, or even preloaded wallpaper, then you can change the appearance of the home and lock screens.

- Press and hold the screen from a home screen

- After that, click Wallpaper
- For available wallpapers, simply click one of the following menus;
  1. Wallpaper services: enable extra features
  2. My wallpapers: select from downloaded and featured wallpapers
  3. Gallery: Select videos and images saved in the Gallery app.
  4. Explore more wallpapers: Go to galaxy themes to locate and download more wallpapers
  5. Apply Dark mode to Wallpaper: enable to apply Dark mode to your wallpaper
- To select the specific video or image, simply click on it
- Once you have selected your specific image, simply select the screen or screens you want it to display.
- The lock screen can only display videos and multiple images
- Depending on which screens are applicable, click Set on Lock screen, Set on Home screen, or set on Home and lock screens.
1. Enable sync my edits if you want to apply wallpaper to both lock and home screen. Besides, this feature

ensures that edits applied to the specific wallpaper will show on the lock and home screens.

Themes

Set a theme to show on your wallpapers, home screen, lock screen, and even app icons.

- Press and hold a screen from a home screen
- After that, click Themes to customize
- Also, click a theme to not only preview but download it to My themes.
- Then, click My page >Themesto see downloaded themes.
- Afterward, click a theme
- Finally, click Apply to apply the chosen theme.

ICONS

To replace the default icons, simply apply another icon sets;

- Press and hold the screen from a home screen
- To customize, simply click Themes icon
- To preview and download an icon to My icons, simply click on it.

- After that, click My page > Icons to access the downloaded icons
- Then, click an icon
- Finally, click Apply to apply the chosen icon set.

Home screen settings

Customize not only your apps screen but your home screens.

- Press and hold the screen from a home screen
- After that, click Home screen settings to customize
1. Lock Home screen layout: Items on the home screen will be prevented from being repositioned or removed
2. Home screen layout: Set your device to have only a home screen where all apps are found or where your device have separate home and apps screens.
3. App icon badges: to access badges on applications with active notifications, simply enable it by choosing this feature. Besides, you select the style for your badge.
4. Home screen grid: Determine how icons are placed on the home screen by choosing a layout.
5. Apps button: For easy access to the Apps screen, simply add a button to the home screen.

6. Apps screen grid: Determine how icons are placed on the apps screen by selecting a layout.

Add Apps to Home screen

Newly-download apps are automatically added to the home screen.

- For notification panel, simply swipe down: To access the notification panel, simply enable the feature by swiping down on the Home screen.
- Hide applications: From the home or app screens, simply select applications to hide. To restore hidden apps, simply return to this screen. Apps that are hidden are still installed and you can find it in the Finder searches.
- Rotate to landscape mode: When your device's orientation changes from portrait to landscape, simply rotate the home screen automatically.
- About Home screen: Access the version information.

Easy mode

Larger text and icons are displayed in the easy mode layout which makes sure you experience a more straightforward visual. You can change between the simpler layout and the default screen layout.

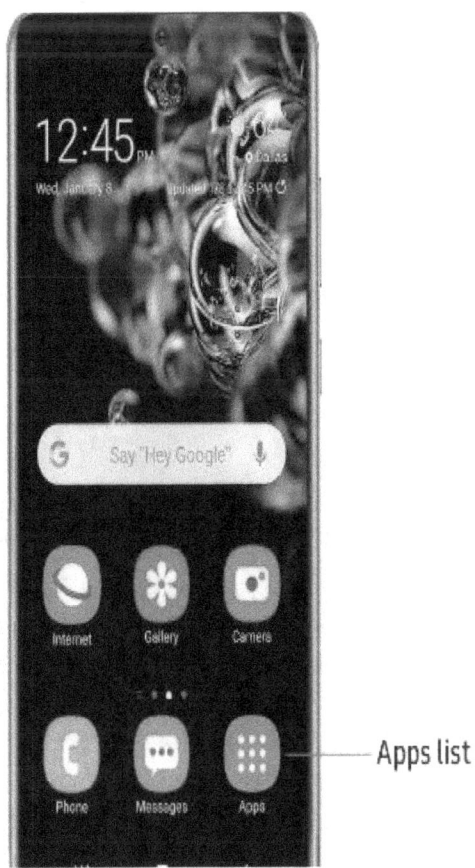

Apps list

- Click Display > Easy mode from settings
- To enable this feature, simply click on it. Then, you will access the options listed below;
  1. High contrast keyboard: Select a keyboard with high contrast colors.

2. Press and hold delay: Set the length for a continuous touch to be recognized as a press and hold.

Configure display options for the status bar.

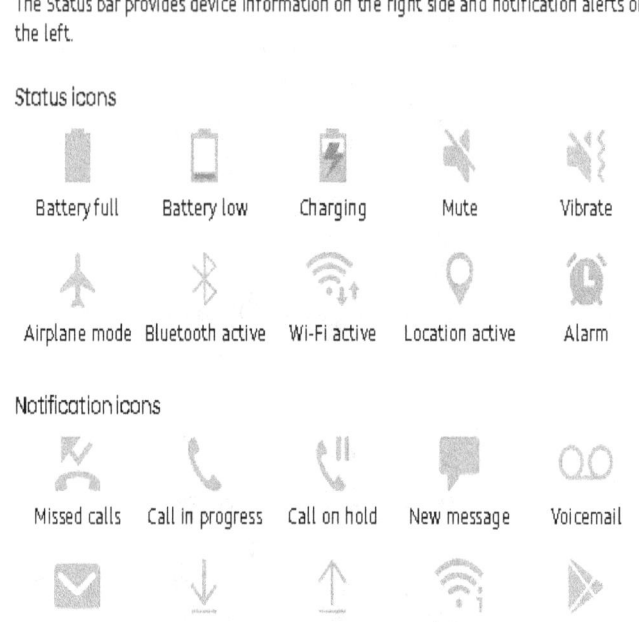

Status bar

The Status bar provides device information on the right side and notification alerts on the left.

Status icons

| Battery full | Battery low | Charging | Mute | Vibrate |

| Airplane mode | Bluetooth active | Wi-Fi active | Location active | Alarm |

Notification icons

| Missed calls | Call in progress | Call on hold | New message | Voicemail |

| New email | Download | Upload | Wi-Fi available | App update |

- Click More options > Status bar from quick settings to access the following options;

1. Show notification percentage: On the status bar, simply show the percentage of your battery alongside the battery icon.

2. Show notifications icons: On the status bar, simply select how to show notification icons.

Quick settings

With quick settings, the notification panel ensures you have quick access to device functions.

- To display the notification panel, simply drag the status bar down
- After that, drag View all downward
- To turn it on or off, simply click a quick setting icon
- To open the setting, simply press and hold a quick setting icon
- To search the device, simply click Finder search
- To power off, restart, or even for emergency mode options, simply click the Power off button.
- To quickly access the device's settings menu, simply click Open settings
- To reorder quick settings or to switch the button layout, simply click More options
- To close quick settings, simply drag View all upward.

Bixby

A virtual assistant that learns, evolves, and even adapts to an individual is often regarded as Bixby. Your routines are learnt and it also helps you set reminders regardless of the location and time. Besides, this feature is built in to apps you prefer.

For more information, simply visit Samsung.com/us/support/owners/app/Bixby

- Touch and hold the sidekey from a home screen

Tip: Once you visit the apps list, you can easily access Bixby.

Bixby Routines

You can access information or change device settings regardless of what you are doing and where you are if you use Bixby.

- Click Advanced features > Bixby routines from settings.

Bixby vision

Bixby gives you a deeper understanding of what you see since it is integrated with your gallery, camera, and internet apps. It offers contextual icons for QR code detection, translation, shopping, or landmark recognition.

It also helps you comprehend what you see since the CameraBixby vision is available on the camera viewfinder.

- Click More > Bixby Vision from camera
- After that, follow the prompts.

Gallery

Images saved in the gallery app can also be used by Bixby vision.

- Click an image from gallery to view it
- Then, click Bixby vision and follow the prompts.

Internet

You can find out more about a picture you find in the internet app as long as you use Bixby vision.

- Press and hold a picture from the internet till a pop-up menu is showed
- Click Bixby vision
- After that, follow the prompts.

Digital wellbeing and parental controls

By getting a daily view of how frequently you use apps, how often you check your device, and the number of notifications you receive, you can easily monitor and manage your digital habits. Also, you can set your device to help you wind down before sleeping in the night.

- Click Digital wellbeing and parental controls from settings to access the following features:
    1. Unlocks: Click to know the number of times each app has been launched today
    2. Notifications: Click to know the number of times notifications have been received from each app today.

AOD themes

For Always on Display, always apply custom themes.

- Press and hold the screen from a home screen
- After that, click Themes > AODs
- To preview and download it to My Always on Display, simply click an AOD

- To see downloaded AODs, simply click My page > AODs
- Click an AOD
- Finally, click Apply

Biometric security

If you want to securely unlock your device and also log in to accounts, simply use biometrics.

Face recognition

You can unlock your screen by enabling face recognition. You must set a PIN, pattern, or even password before you can use your face to unlock your device.

- Face recognition does not offer strong protection unlike the PIN, password, or pattern. Someone or something that looks like your image can unlock your device.
- The face recognition may be affected by some conditions like hats, putting on glasses, beards, or heavy make-up.
- Make sure that when you are registering your face, you are in a well-lit area and the camera lens is clean.

1. Click Biometrics and security > Face recognition from settings

2. To register your face, simply follow the prompts. Face recognition management customize how face recognition works.

- Click Biometrics and security > Face recognition from settings

1. Remove face data: get rid of existing faces

2. Face unlock: Disable or enable face security

3. Add alternative look: Add an alternative appearance to enhance face recognition.

- Stay on lock screen: Stay on lock screen if you use face recognition to unlock your device. Unless, you swipe down.

- Faster recognition: For faster recognition, simply turn it on. To increase security and make it more difficult to unlock using a video or image of your likeness, simply turn it off.

- Require open eyes: When your eyes are open, facial recognition will recognize your face.

- Brighten screen: For your face to be recognized in dark conditions, simply increase the screen brightness temporarily.
- Samsung pass: Use face recognition to access your online accounts.
- About unlocking with biometrics: Acquire more knowledge about securing your device with biometrics.

Fingerprint scanner

You can also enter passwords in certain apps by using fingerpint recognition. Besides, you can verify your identity when logging in to your Samsung account if you use the fingerprint feature. You must set a PIN, pattern, or password if you want to use your fingerprint to unlock your device.

- Click Biometrics and security > Fingerprints from settings
- To register your fingerprint, simply follow the prompts. Fingerprint management Add, rename, and delete fingerprints.
- Click Biometrics and security > Fingerprint from settings to access the following options;

1. At the top of this list, you will see the list of registered fingerprints. You can rename or even remove it by clicking a fingerprint.
2. Check added fingerprints: To know if your fingerprint has been registered, simply scan it.
3. Add fingerprint: to register another fingerprint, simply follow the prompts.

Fingerprint verification settings

To verify your identity in supported actions and apps, simply use fingerprint recognition.

- Click Biometrics and security > Fingerprints from settings
  1. Fingerprint unlock: when unlocking your device, simply use your fingerprint for identification
  2. Show icon when screen is off: When the screen is off, simply show the fingerprint icon
  3. Samsung pass: when using supported apps, use your fingerprint for identification
  4. Samsung pay: To make payments quickly and securely, simply use the fingerprints

5. About unlocking with biometrics: Go through the details on requirements every biometric security feature possess for using the PIN, pattern, or password.

Link your device to your computer

- Click Advanced features > Link to windows from settings
- To enable this feature, simply click on it
- To connect your device to a computer, simply follow the prompts.

TIP: you can also head to quick settings menu to enable this feature.

Samsung Dex for PC

For an enhanced and multitasking experience, simply connect your device to a computer.

- Use your computer and device apps side-by-side.
- Between the computer and your device, simply share the mouse, keyboard, and screen.
- Use Dex to send texts or make phone calls.

For more information, visit Samsung.com/us/explore/dex

Set up Dex on your computer

- Use a Standard USB-C cable to connect your device to a computer

- On your device for downloading and installing the Dex for PC software on your computer, simply follow the instructions.

Multi window

Use multiple apps at the same time to practice multitask, although these apps must be supported by Multi window and can be displayed together on a split screen. Switching between the apps and adjusting the size of their windows is feasible with this feature.

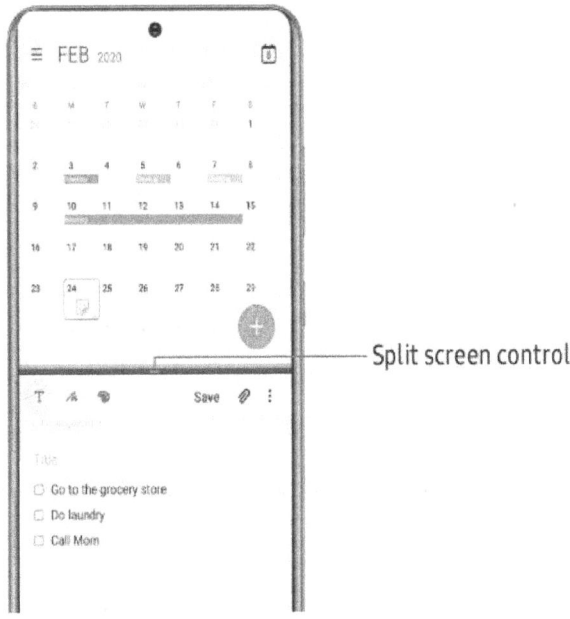

Split screen control

- Click Recent apps from any screen
- Click the app icon
- Then, click Open in split screen view
- In the other window, click an app to add it to the split screen view
- To adjust the window size, simply drag the middle of the window border.

Apps panel

Up to ten apps in two columns can be added to the Apps panel.

- Drag the edge handle from any screen to the center of the screen. Swipe till you notice the apps panel.
- To open an app or app pair, simply click on it. Configuring apps panel requires you to;

1. Drag the edge handle from any screen to the center of the screen. Swipe till you notice the apps panel.
2. To add other applications to the apps panel, simply click Add apps to folder.

- Find an app on the left side of the screen, then click it to add apps to the apps panel.
- Click Create app pair to create a shortcut for two apps to launch in Multi window.
- Drag an app from the left side of the screen at the edge of the app in the columns on the right to create a folder shortcut.
- Drag each app to the desired location to change the order of the apps on the panel.
- Click Remove to remove an app

3. To save changes, simply click Back

Smart Select

This feature (smart select) captures a location in the screen as an animation or even image and that you can pin or share to the screen.

1. Drag the edge handle from any screen to the centre of the screen. Swipe till you notice the apps panel.
2. Use the following options by clicking a smart select tool;

- Pin to screen: Capture an area and pin it to the screen
- Rectangle: Take a rectangular area of the screen
- Animation: record activity on the screen as an animated GIF
- Oval: Take an oval area of the screen

Configure Edge panels

Customizing edge panels requires you to;

- Click Settings from the edge screen
- Enable the feature by clicking on it. Then, you will access the following options;
  1. Edit (if available): configure individual panels
  2. Checkbox: Disable or enable each panel

3. Search: Look for panels that could be installed or even available to install.

Edge lighting

When you receive calls or notifications, simply set the edge screen to light up. It ensures the alerts are visible when the screen is face-down.

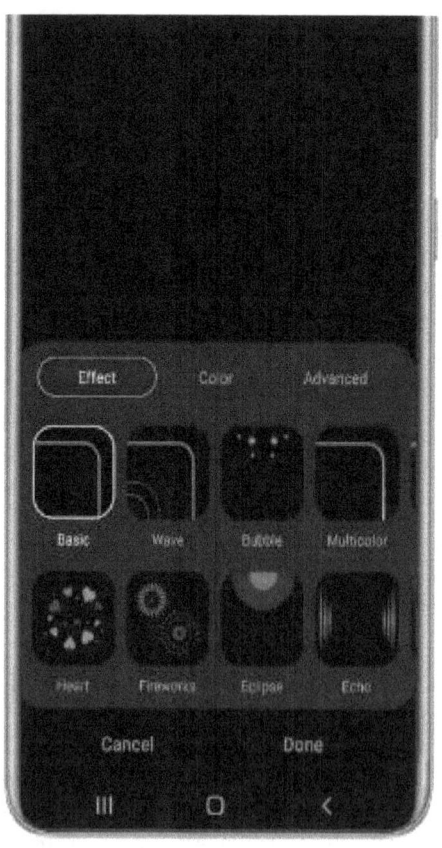

- Click Display > Edge screen > Edge lighting from settings
- After that, click the feature to enable it.

Lightning style

With the edge lightning feature, you can customize the transparency, width, and color of the screen.

- Click Display > Edge screen > Edge lightning from settings
- To customize, simply click Lightning style
  1. Color: Select a custom or preset color, and enable app colors. To configure a custom lightning for specific text that shows in the notification titles, simply click Add keyword.
  2. Width: to adjust the width of the edge lightning, simply drag the slider
  3. Transparency: to adjust the transparency of the edge lightning, simply drag the slider
  4. Duration: to adjust how long or short the edge lightning displays, simply drag the slider.
- When you finish, simply click done.

Select apps

Select the apps that will help in activating edge lighting.

- Click Display > Edge screen > Edge lighting from settings
- To select the specific apps that will help in activating edge lighting especially when you receive a notification, simply click Choose apps.

Enter Text

You can use a keyboard or even your voice to enter text.

Configure the Samsung keyboard

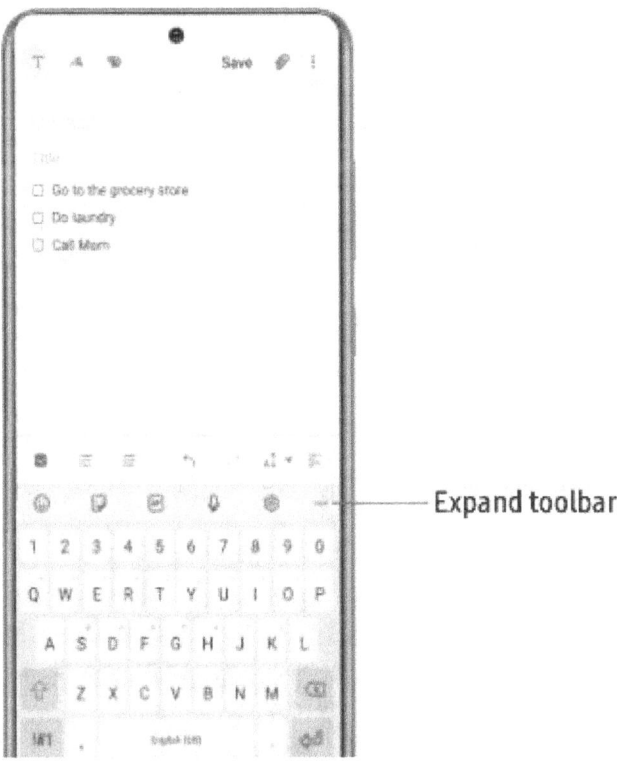

Expand toolbar

For the Samsung keyboard, simply set customized options.

- Click Settings from the Samsung keyboard to access the options listed below;

1. Smart typing: To prevent common typing mistakes, simply use predictive text and auto-correction features. Swipe between letters to type.

2. Languages and types: Select which languages are available on the keyboard and set the keyboard type.

Swipe the space bar right or left to change between languages.

3. Style and layout: customize the function and appearance of the keyboard.

4. Reset to default settings: Clear personalized data and return keyboard to original settings

5. Swipe, press, and feedback: Customize feedback and gestures

6. About Samsung keyboard: Have access to legal information and view version for the Samsung keyboard.

Use Samsung voice input

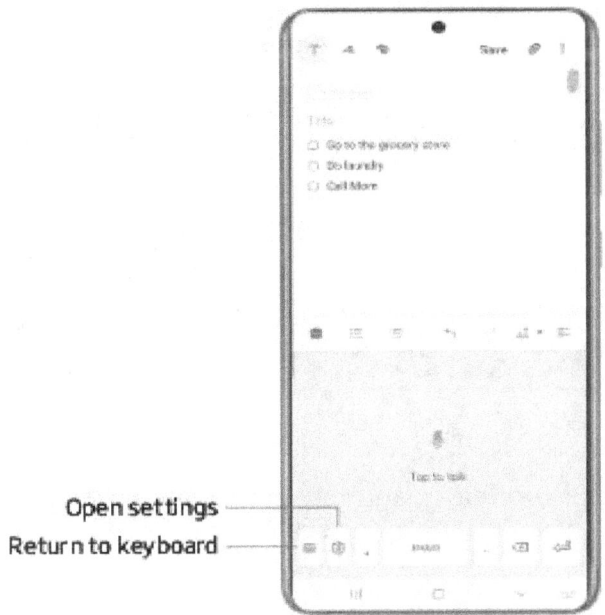

Open settings
Return to keyboard

Use voice to enter text rather than typing.

- Click Voice input from the Samsung keyboard
- After that, speak your text.

Camera and Gallery

Use the camera app to capture high-quality images and videos. The gallery app stores images and videos. It is also where you access them and edit them.

Enjoy a full kit of pro lenses and pro-grade video models and settings.

Click Camera from apps

TIP: Quickly touch the side key twice if quick launch is enabled.

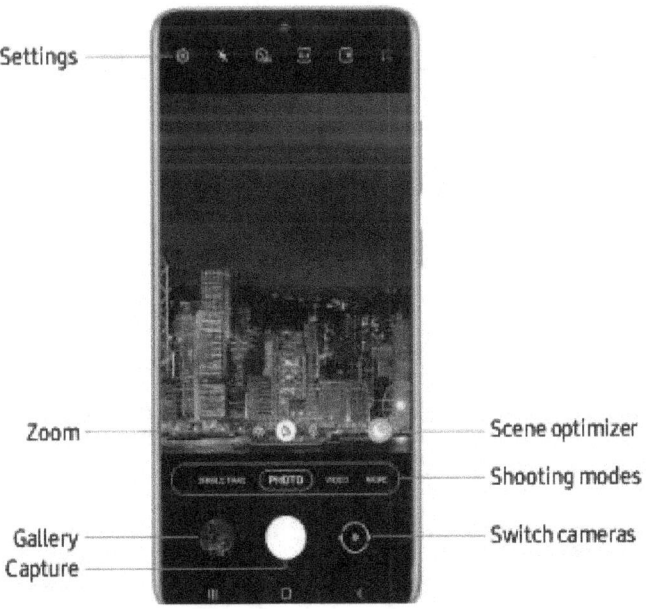

Settings

Zoom

Gallery
Capture

Scene optimizer

Shooting modes

Switch cameras

Navigate the camera screen

Use your device's front and rear cameras to take stunning images.

- Set up your shot from camera with the following features;

1. Where you want the camera to concentrate, simply click the screen

2. A brightness scale pops-up when you click the screen. Adjust the brightness by dragging the cycle.

3. Swipe the screen down or up to quickly change between the rear and front cameras

4. Swipe the screen right or left to switch to a different shooting mode.

5. Click settings to change camera settings

- Click capture. To determine the perfect mode for your images or even select from numerous shooting modes, simply configure shooting mode.

   1. Swipe the screen left or right from camera to change shooting modes.

   2. Video: To determine the perfect settings for videos, simply allow the camera

   3. Photo: To determine the perfect settings for images, simply allow the camera

   4. Single take: To capture the scene in a series of short clips and images, simply click the shutter button

   5. More: other available shooting modes can be selected. To drag modes into or out of the Modes tray at the bottom of the screen, simply Click Edit.

      – Pro: while taking images, manually adjust the

exposure value, iSO sensitivity, color tone, and white balance.

6. Night: Without using the flash, use this to take images in low-light conditions.

7. Panorama: By taking images in either vertical or horizontal direction, you can create a linear image.
   – Food: capture images that emphasize the clear colors of food.

8. Live focus video: use adjustable depths of focus to record artistic videos

9. Live focus: Adjust the depth of field by capturing artistic images

10. Pro video: while taking images, manually adjust the exposure value, iSO sensitivity, color tone, and white balance.

11. Super slow-mo: To video in high quality slow motion, simply record videos at an extremely high frame rate. After recording it, you can play one part of each video in slow motion.

12. Hyper lapse: By recording at several frame rates, you are creating a time lapse. Depending on the scene being recorded and the movement of the device, the frame rate is adjusted.

13. Slow motion: To view in slow motion, simply record videos at a high frame rate.

Live Focus

You can also make your pictures have an added interactive focus effects.

- Swipe to More from camera
- After that, click Live focus
- Also, click Live Focus effect
- Select an effect
- Finally, fine-tune the effect by dragging the slider.

Scene optimizer

If you want to capture beautiful photos, then try to automatically adjust contrast, exposure, white balance, and more based on what you noticed in the camera frame.

- Swipe to photo from camera
- After that, click Scene optimizer

Note: when you use the rear camera, you have access to the scene optimizer. When you take nature photos or when you take photos in a dark setting, then the scene optimizer icon will change.

Record Videos

Use your device to record high-quality videos.

- Swipe left or right from camera to switch the shooting mode to video
- After that, start recording a video by clicking Capture
- Click Capture to capture an image while recording
- Click Pause to temporarily stop recording. Click Resume to continue recording.
- When you are finished recording, simply click Stop.

Live Focus Video

Once you apply background blurs and other unique effects to your video, you are directly creating professional-looking films. You cannot use zoom, super steady, or zoom-in mic on this feature.

- Swipe to More from camera
- After that, click Live focus video
- Also, click Live focus effect

- Select an effect
- Then, drag the slider to fine-tune the effect
- Finally, begin recording by clicking Capture.

Zoom-in mic

As you zoom in on an audio source, try to increase the volume of the sound being recorded and also reduce the background noise. You cannot use this feature with Super steady, live focus video, or with the front camera.

- Click Setting from camera
- Click Advanced recording options > Zoom-in-mic
- After that, click to enable
- To go back to the main camera screen, simply click Back
- To change the shooting mode to video, simply Swipe
- To start recording, simply click Capture
- To zoom in or out on the audio source, simply bring your fingers apart or together on the screen. The level of amplification being applied is usually indicated by the microphone icon.

Super Slow-mo

If you wish to view videos in slow motion, then record it at a high-frame rate.

- Swipe to More from camera
- After that, click Super slow-mo
- Click Super Slow-Moto record

TIP: To get the best results, simply hold your device steady.

Super steady

If you want your video to have a smooth, professional appearance, and a heavy motion situation, then use this feature because it applies advanced stabilization algorithms. You cannot use this feature with Live focus video, Zoom-in mic, or even the front camera.

- Swipe to change the shooting mode to video when you arrive at the camera app
- After that, click Super Steady
- Finally. Start recording by clicking Capture.

Camera Settings

Go to the main camera screen to use the icons and the settings menu to configure the settings of your camera.

- Click Setting from camera
- Then, access the options listed below;

1. Scene optimizer: If you want the subject matter to match, then automatically adjust the color settings of your images.

2. Shot suggestions: If you want to select the best shooting mode, then get tips from this feature.

3. Smart selfie angle: When there are more than two individuals in the frame, automatically change to a wide-angle selfie.

4. Scan QR codes: When you use the camera, you can automatically detect QR codes.

5. Swipe Shutter button to edge to: When you swipe the shutter to the nearest edge, simply create a GIF or elect to either take a burst shot.

6. Save options: Select other saving options along with file formats.

7. HEIF pictures (Photo): To save space, simply save images as high efficiency. This format may not be supported by some sharing sites.

8. Save RAW copies: Pictures taken in pro mode can be saved in RAW and JPEDG copies.

9. Ultra wide shape correction: Pictures taken with the ultra wide lens videos, simply automatically correct distortion.

10. Rear video size: Choose a resolution. More memory is required if you are choosing a higher resolution for higher quality.

11. Front video size; Choose a resolution. More memory is required if you are choosing a higher resolution for higher quality.

12. Advanced recording options: Use the advanced recording formats to enhance your videos

13. High efficiency video: Save space by recording videos in HEVC format. This playback of this format may not be supported by other device or sharing sites.

14. –HDR10+ Video: record in HDR10+ to optimize videos. HDR10+ videos is supported by playback devices.

15. Zoom-in-mic: while recording videos, simply match the camera zoom with the mic zoom.

16. Video stabilization: To keep the focus steady when the camera is in motion, simply activate anti-shake. Ensure the features are useful.

17. Auto HDR: In bright and dark areas of your shots, simply capture more detail.

18. Selfie tone: Add a cool tint or warm tint to your selfies.

19. Tracking auto-focus: Keep a moving subject in focus.

20. Pictures as previewed: As selfies appear in the preview, simply save them without flipping them.

21. Grid lines: compose a video or picture as they display viewfinder grid lines.

22. Location tags: On your picture, simply attach a GPS location tag.

23. Press volume key to: To take pictures, control system volume, or record video, simply use the volume key.

24. Voice control: speak key words to take pictures.

25. Floating shutter button: An additional shutter button you can move anywhere on the screen, simply add it.

26. Show palm: use your palm facing the camera to take quick pictures.

27. Storage location: Choose a memory location. View storage location without having the memory card.

28. Shutter sound: When taking a picture, simply play a tone.

29. Reset settings: Reset the settings of the camera

30. About Camera: Have access to the app and software information.

Gallery

If you want to access all the visual media stored on your device, simply head over to the Gallery. When you access the gallery, you should be able to view, manage, and edit videos and pictures.

Sort images into custom albums

View pictures and videos

Customize collections of pictures and videos

Share pictures and videos

- Click Gallery from Apps

View pictures

You can view pictures in the Gallery app by;

- Clicking Pictures from Gallery
- After that, click a picture to view it. To view other videos or pictures, simply swipe right or left.
- Click Bixby vision to use Bixby vision on the current image.

Visit Bixby to get more details.

- Click favorite to mark the picture as a favorite
- Click More option to access the options listed below;
  1. Set as wallpaper: Set the image as wallpaper
  2. Details: View and edit information about the image.
  3. Set as Always on Display image: For the Always on Display, simply set the image as the background image.
  4. Send as live message: To draw an animation on a picture and even send it, simply use Live message.
  5. Move to Secure Folder: Place the image in a secure folder.
  6. Print: Send the image to a printer that is already connected.

Edit pictures

Use the Gallery's editing tool to enhance your images.

- Click Pictures from Gallery
- To view the specific picture, simply click on it.
- Have access to the following options once you click Edit;

1. Tone: Adjust the exposure, brightness, contrast, and more.

2. Filter: Add color effects.

3. Transform: Crop, rotate, flip, or make other changes to the whole appearance of the image.

4. Sticker: Overlay animated or illustrated stickers.

5. Draw: Add hand drawn or text content.

- When you are finished, simply click Save.

Play Video

Videos stored on your device, simply access them. You can view video details and even save them as favorites.

- Click Pictures from gallery
- To view the specific video, simply click on it. To view other pictures or videos, simply swipe right or left.
- Click Favorite to mark the video as a favorite. Just beneath the Albums tab, you will see the video added to Favorites.
- Click More options to access the options listed below;

1. Move to secure Folder: Add the video to your secure folder

2. Details: View and edit information about the video

3. Send as live message: To draw an animation and even share it, simply use Live message

4. Set as wallpaper: On the lock screen, simply set the video as wallpaper.

- Finally, play the video by clicking Play Video.

Video enhancer

To enjoy brighter and even clearer colors, simply enhance the image quality of your videos.

- Click Advanced features > Video enhancer from settings
- After that, click the feature to enable it.

Edit Video

Videos stored on your S21 device, simply edit them by;

- Clicking Pictures from gallery
- After that, view a specific video by clicking on it.
- Use the following tools when you click the Edit button;
  1. Draw: Draw on your video
  2. Trim: Cut segment of the video
  3. Speed: Adjust the speed of play

4. Filters: Add visual effect to the video
5. Audio: Add background music to the video and also adjust the volume levels
6. Portrait: Enhance the eyes, skin tones, and other facial features
7. Text: Add text to your videos
8. Rotate: Rotate the video clockwise

- Click Save and when prompted, simply confirm.

Share picture and Videos

From the gallery app, simply share pictures and videos.

- Click Pictures from gallery
- After that, click More options > Share
- Also, click Pictures and videos to choose them
- Click the Share button
- Then, select an app or connection to use for sharing your selection.
- Finally, follow the prompts.

Delete pictures and videos

Pictures and videos stored on your device can also be deleted.

- Click More options > Edit from gallery
- To select a specific picture or video, simply click on it.
- You can also head over to the top of the screen and click the Allcheckbox to choose all picture and videos
- Click Delete
- Finally, when prompted; simply confirm.

Create Movie

From the camera app, simply share videos and pictures by using the video effects and music to create a slideshow of your content.

- Click Create move from Gallery
- To add a picture or video to the movie, simply click on it.
- Click Create movie
- After that, select either Self-edited (custom slideshow) or Highlight reel (Automatic slideshow).
- You will access the options listed below;
1. Title: Add a description and a title to your movie
2. Duration: Adjust the run time of the whole movie. Keep note it is only highlight reel.

3. Clips: Access and edit each picture or video in your movie.

4. Transition effect: Customize the transitions between each clip (Self-edited only) to add visual interest to your movie.

5. Add: From the gallery (self-edited only), simply incorporate extra clips.

6. Share: Send your movie to family and friends.

- Clive save.

Take a screenshot

Take a picture of your screen. You will create a screenshot album in the gallery app.

- Touch and hold the side and volume down keys from any screen. You can also capture a screenshot by using the palm swipe.

By swiping the edge of your hand across it, you can capture an image of the screen even if you keep in contact with the screen.

- Click Advanced features > Motions and gestures > Palm swipe from settings to capture.

- Click this feature to enable it.

Screenshot settings

Control screen recorder and screenshot settings.

- Click Advanced features > Screenshots and screen recorder from settings
1. Screenshot format: If you want the screenshots to be saved as PNG files or JPG files, simply choose this feature.
2. Screenshot toolbar: After you capture a screenshot, simply choose this feature to show extra options
3. Delete shared screenshots: You can automatically delete screenshots once it is shared through the screenshot toolbar.

Screen recorder

On your device, simply record activities, write notes, and record a video overlay of yourself to share with family and friends by using the camera.

- Click Screen recorder from quick settings to start recording.

- To draw on the screen, simply click Draw
- Click Selfie video to add a recording from your front camera.
- To finish recording, simply click Stop. In the Gallery, you will see saved screen recording album.

For the screen recorder, simply set the screen recorder settings to control the quality settings and sound.

- Click Advanced features > Screenshots and screen recorder > Screen recorder settings from settings
  1. Selfie video size: To set the size of the video overlay, simply drag the slider
  2. Video quality: choose a resolution. More memory is required for higher quality.

Apps

Using App

Download Apps

All preloaded and downloaded apps are displayed in the apps list. You can download apps from the galaxy store and the Google play-store.

- Swipe the screen upward from a home screen to access the apps list.

Uninstall or disable apps

You can remove installed apps on your device. Some of the preloaded apps available on your device by default can only be disabled. Besides, disabled app are usually hidden from the apps list and turned off.

- Press and hold a specific app from apps
- After that, click Uninstall or Disable.

Create and Use Folders

You can organize app shortcuts to the apps list by making folders.

- Press and hold a specific app shortcut from apps
- After that, drag it on top of another app shortcut till it is highlighted.
- To create a folder, simply release the app shortcut.
  1. Folder name: Name the folder
  2. Palette: Change the folder color.

3. Add apps: In the folder, simply place more apps. To select then, click apps and also click Done.

- To close the folder, simply click Back.

You can also copy a folder to a home screen.

- Press and hold a folder from apps
- After that, click Add to Home

Delete a Folder

The app shortcuts of a folder will return to the apps list when it gets deleted.

- Press and hold a folder from apps to delete
- After that, click Delete folder
- Finally, when prompted; simply confirm.

Samsung applications

The applications listed below are either pre-downloaded on your phone or are downloaded as you set up your device.

Galaxy Essentials

The Galaxy Essentials are a variety of special apps available on Samsung devices. To download these premium apps, navigate to Apps > More Options > Galaxy Essentials.

AR Zone

The AR Zone houses all Augmented Reality features. Visit the AR Zone to know more. Navigate to Apps > Samsung folder > AR Zone.

Bixby

Depending on how you interact with your device, Bixby suggests and displays contents that stem from your interactions. Visit Bixby to know more. Navigate to Apps > Samsung Folder > Bixby.

Galaxy Store

You can download exclusive and premium apps from the Galaxy Store on your Samsung Galaxy devices. Navigate to Apps > Galaxy Store.

Galaxy Wearable

With this app, you can pair your Samsung Watch to your device. Visit samsung.com/us/support/owners/app/galaxy-wearable-watch to know more. Navigate to Apps > Galaxy Wearable.

Game Launcher

Use the game launcher to automatically arrange all your games in one location. Visit samsung.com/us/support/owners/app/game-launcher to know more. Navigate to Apps > Game Launcher.

Note that if you can't find Game Launcher in the list of apps, an alternative way to find it is to navigate to Advanced features > Game Launcher.

Samsung Global Goals

The ads from this app showcase Samsung's Global Goals initiative and her contributions and donations supporting these initiatives. Navigate to Apps > Samsung Global Goals.

Samsung Members

This app gives you more than you can do with your device. You can enjoy exclusive content and DIY support features. This app comes preloaded or you can download it from Google Play Store or Galaxy Store. Navigate to Samsung Folder > Samsung Members.

Internet

The Samsung Internet is a reliable, swift and simple, web browser. Its security features include the Content Blocker, Secret Mode, and Biometric Web Login. Navigate to Apps > Internet.

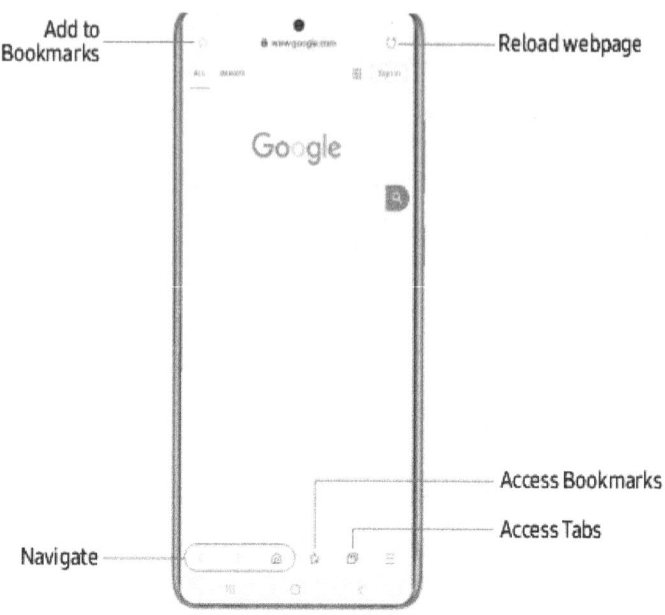

Add to Bookmarks

Reload webpage

Navigate

Access Bookmarks

Access Tabs

TIP Visit samsung.com/us/support/owners/app/samsung-internetfor more information

Browser tabs

View different web pages simultaneously using the tabs.
Navigate to Internet > Tabs > New tab.

To close a tab, navigate to Tabs> Close tab.

Bookmarks

Your browsing history saved pages, and bookmarks are saved on the Bookmarks page.

To open a bookmark, go to the bookmarks page and launch any web page.
Navigate to Internet > Bookmarks and tap on any bookmark entry.

Save a web page

When you save a web page, its contents are stored on your device to allow you offline access. Navigate to Tools > Add page to > Saved Pages.

To see the saved pages, navigate to Tools > Saved Pages.

To see web pages recently visited navigate to Internet > Tools > History.

To delete your browsing history, navigate to More options > Clear history.

Share pages

You can share your visited web pages with your contacts. Navigate to Internet > Tools > Share, then follow the instructions.

Messages

Use the Messages app to keep in touch and connect with your contacts. With it, you can send an emoji, share photos or simply say a quick hello. Navigate to Apps > Messages > Compose a new message.

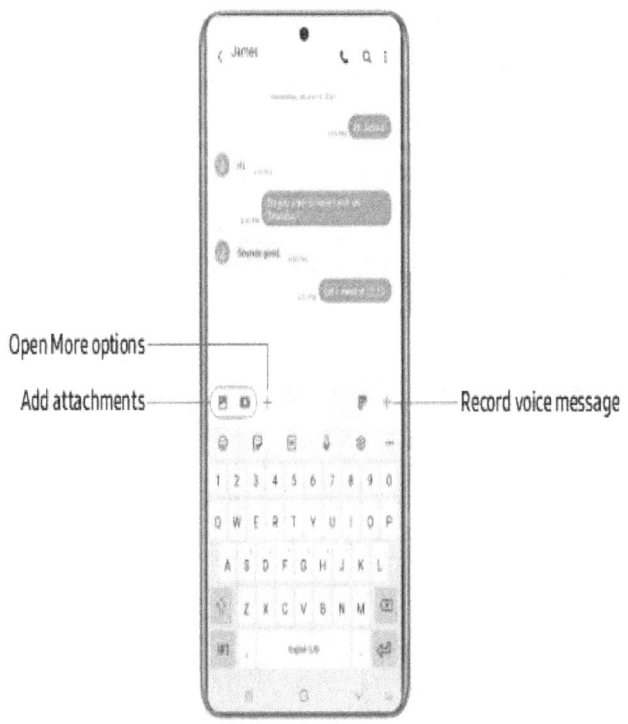

Open More options

Add attachments

Record voice message

Message search

Make use of the search feature to quickly find any message.

To do this, navigate to Messages > Search. Then input the keyword in the search box and tap Enter.

My Files

Manage and view the files that are saved on your device. These include music, recordings, images, and videos. Navigate to Apps > Samsung Folder > My Files

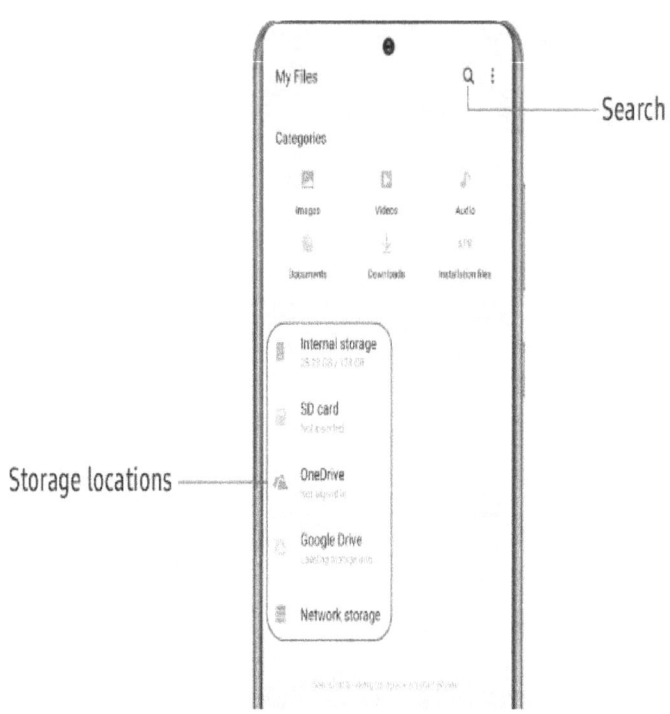

File groups

The files that are saved in your device are arranged into the following groups:

•Categories: See the files based on their type.

•Recent files: See the files recently accessed.

•Analyze storage: See the items filling up your storage space.

•Storage: See the files that are stored on your SD card, device memory, or cloud accounts. The cloud accounts may differ depending on the service that you signed up for.

My Files options: You can make use of the My Files options to edit, search, clear file history, and more.

You can find the following options when you access My Files:

•Search: Search for a folder or file.

•More options:

–Clear Recent files list: Remove the recently accessed files list. You can only view this option when you have opened a file via My Files.

–Analyze storage: See what's taking up your storage space.

–Trash: Permanently delete or restore the files that you have deleted.

–Settings: View the app settings.

Phone

You can perform more than just phone calls using the Phone app. Relish the advanced call features. For more info, visit your carrier. To access this app, navigate to Home screen > Phone.

Access voicemail

Make a video call

Make a call

Calls

Using the Phone app, you can answer and make calls through Contacts, from the Home screen, from the Recents tab, and more.

Make a call

From the Home screen, you can make and answer calls using your phone. To do this navigate to Phone, input a number on the dial pad, and press Call. Tap the Keypad if it is not automatically displayed.

Enable Swipe to call

To make a call, you can also swipe a number or contact to the right to place the call. To enable this, navigate to Settings > Advanced features > Motions and gestures > Swipe to call or send messages.

Make a call from Recents

The Call log stored the log of all missed, incoming and outgoing calls.

To display the recent call list navigate to Phone >Recents

Make a call from Contacts

Call a contact via the Contacts app.

In the Contact app, call a contact by swiping your finger on a contact to the right.

Answer a call

When you receive a call, the caller's name and phone number are displayed on the phone as it rings. If you are making use of an app, a popup screen is displayed containing the caller's contact IDs.

To answer a call, drag Answer on the incoming call screen to the right. While using an app, simply tap Answer to pick the call.

Decline a call

You can decide to decline an incoming call.

To reject a call and send it to voicemail, drag Decline on the incoming call screen to the left.

While using an app, tap Decline on the pop-up screen to decline the call.

Decline with a message

You can make use of a text message response to decline an incoming call.

Drag the Send message option on the incoming call screen upward and choose a message.

While using an app, tap Send message and choose a message.

End a call

Just tap End when you are ready to end the call.

Actions while on a call

While on a call you can switch to a speaker or headset, adjust the call volume, or multitask.

To adjust the volume, press the volume keys.

Switch to headset or speaker

Make use of a Bluetooth® headset or speaker to listen to the call. To do this tap Bluetooth to make use of a Bluetooth headset or speaker to hear the caller via the speaker.

Optional calling services

The following call services are supported if they are available on your service plan.

Place a multi-party call

Depending on your service plan support, you can make another call while another call is ongoing.

To do this, tap Add call while there is an ongoing call to dial the second number. Input the new number and press Call.

When the call has been answered, to can:

•Switch between the calls by tapping Swap.

•Activate the multi-conferencing feature by tapping Merge. This allows you to jest the callers at once.

Video calls

To make Video calls:

Navigate to the Phone app, input a phone number, and tap Video call.

Be aware that video calling is not supported on all devices. The receiver will have the option to accept the call as either a video call or a voice call.

Wi-Fi calling

If this is supported by your mobile network, you can make calls over a Wi-Fi network. For more details, contact your carrier. Navigate to Phone > More options > Settings > Wi-Fi calling and follow the instructions to configure it.

Real-Time Text (RTT)

While on a call, you can type with the other individual in real-time. You can make use of RTT only if the other person uses a phone that supports it or is connected to the teletypewriter (TTY). If there is an incoming RTT call, the RTT icon will appear. To view this navigate to Phone > More options > Settings > Real-time text.

The following options are available after tapping Real-Time Text:

•Always visible: This displays the RTT call button during calls and on the keypad.

•Use an external TTY keyboard: When an external TTY keyboard is connected, this option hides the RTT keyboard.

•TTY mode: Select the TTY mode best suited for the keyboard being used.

Be informed that you have to be connected to a 4G LTE or Wi-Fi network before you can make use of RTT.

Samsung Pay

With the Samsung Pay™, you can make payments using your device. It can be used in almost every place where you can tap or swipe your credit card. To do this, you must have a

Samsung account. Visit

samsung.com/us/support/owners/app/samsung-pay to know

more. To access this, navigate to App > Samsung Pay > Get

started, then follow the instructions.

Be aware that your debit and credit card info is not stored on

the cloud storage for security reasons. While making use of

this service on different devices, you must sign in and confirm

all the payment cards on each of the devices. Depending on

the card issuer, the number of devices may have a limit.

To use Samsung Pay, open the app and hold it over the card

reader of the store. Then do the following:

1. Go to Apps and tap Samsung Pay. Choose a card to pay with

and authorize the payment by inputting your Samsung Pay

PIN or scanning your finger.

2. Hold your phone over the card reader of the store.

•A receipt will be sent to your registered email address upon

complete payment.

Ensure that the NFC feature is activated on your device. See

NFC and Payment for more info.

Simple Pay

You can make use of Simply Pay to access Samsung Pay for either the Always-on display, Lock Screen, or Home screen.

To do this navigate to Apps > Samsung Pay > Menu > Settings > Use Favorite Cards then tap to enable Simple Pay on each screen.

To use Simple Pay:

Swipe up from the bottom of any of the screen interface. This will display Simple Pay and your payment card.

2. Close Simple Pay by dragging the card down.

You can make use of gift cards with Samsung Pay Purchase to redeem and send gift cards from your list of retailers. Visit gift samsung.com/us/samsung-pay/compatible-cards/#bank to see the list of supported gift card merchants and banks.

Secure your information

Samsung Pay works on recent Samsung Galaxy devices and features the latest security. The payments are authorized with your PIN or fingerprint, making use of unique tokens per transaction. This allows for only payments with your consent.

You can make use of the Find My Mobile function to wipe your data off a lost device for utmost protection. See Find My Mobile to know more

Microsoft apps

Enjoy the following Microsoft apps; OneDrive| LinkedIn | Office Mobile | Outlook

Outlook

Outlook features tasks, mail, calendar, contacts, and more. See Add an Outlook account or visit dupport.office.com/en-us/outlook to learn more.

Navigate to Apps > Microsoft folder > Outlook

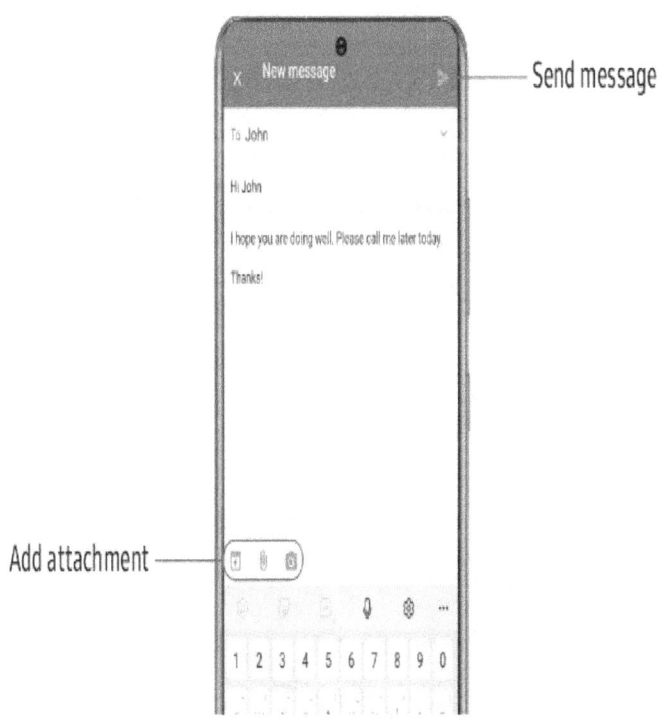

Send message

Add attachment

Access Settings

There are different methods of accessing the settings method of your device. To do this tap and drag the status bar down then tap Settings. Alternatively, navigate to Apps then tap Settings.

Look for Settings

You can look for where Settings is if you are not sure of its location. To do this navigates to Settings then tap Search and input the keyword. Alternatively, tap on any entry to go to that setting.

Connections

Manage the connections between your device, different networks, and devices.

Wi-Fi

Without making use of your mobile data, you can still access the internet by connecting your device via Wi-Fi. Navigate to Settings > Connections > Wi-Fi. Then tap to toggle it on and scan for networks.

If there is a network, tap it and input the password if it is required.

Wi-Fi network manual connection

You can still connect to a network that is not found even after scanning. Simply enter the Wi-Fi information manually. To begin, request the Wi-Fi name and password from the Wi-Fi administrator.

Navigate to Settings > Connections > Wi-Fi and toggle it on. Then tap on Add network located at the list's bottom. Finally, input the Wi-Fi information. This info includes:

•Security: Choose an option for security from the list and input the password if it is required.

•Network name: Enter the exact network name.

•Auto reconnect: Select this option if you want to auto-reconnect anytime the network signal is in range.

•Type of Mac address: Select the Mac address type to be used for the connection.

•Advanced: Include any advanced options like proxy and IP settings and tap save.

You can scan a QR code using the camera of your device to connect to any Wi-Fi network.

Wi-Fi Advanced settings

You can configure and manage your various Wi-Fi hotspots and connections. You can also lookup the addresses of your device's network. To do this, navigate to Settings> Connections > Wi-Fi and toggle it on. Then tap More Options > Wi-Fi.

•Switch to mobile data: If this feature is enabled, your device will switch to mobile data if the Wi-Fi connection is not stable. It switches back when the network becomes stable.

•Automatically turn on Wi-Fi: This feature turns on Wi-Fi in locations that are frequently used.

•Detect suspicious networks: With this, you get notified anytime there is a suspicious activity detected on the Wi-Fi network

•Wi-Fi power saving mode: This reduces the battery consumption by analyzing the Wi-Fi traffic.

•Network notification: If there is an open network detected at the range, you will be notified.

•Manage networks: See the saved Wi-Fi networks and choose whether to forget them or automatically connect to them.

•Wi-Fi control history: See the apps that have recently turned your Wi-Fi off or on.

•Hotspot 2.0: Automatically connect to Wi-Fi networks that are compatible with the Hotspot 2.0.

•Install network certificates: Install the authentication certificates.

•MAC address: See the MAC address of your device. This is usually required when you are connecting to more secured non-configurable networks.

•IP address: See the IP address of your device. This is non-configurable.

Wi-Fi Direct

Wi-Fi Direct makes use of a Wi-Fi connection to share data between devices.

Navigate to Settings > Connection > Wi-Fi, then toggle it on. Then tap More options > Wi-Fi Direct. Finally, tap any device and follow the instructions to connect.

Bluetooth

You can establish a connection between your device and a Bluetooth device such as a Bluetooth-enabled car infotainment system or a Bluetooth headset. Once both devices have been paired, they can exchange information without the need of inputting a password again.

Navigate to Settings > Connections > Bluetooth, then toggle it on. To connect, tap any device and follow the instructions. Note that you can also make use of the Bluetooth feature to share files.

Rename a paired device

To make your paired device easier to be recognized, you can just rename it. To do this navigate to Settings > Connections > Bluetooth and toggle it on.

Next to the device name in Settings, tap it t rename. Then input the new name that you want and tap rename.

Data usage

View the current Wi-Fi and mobile data usage is your device. You can also modify and customize the limits and warnings. Navigate to Settings > Data Usage.

Turn on Data saver

Make use of Data saver to lower the rate of your data consumption by stopping certain apps from receiving or sending background data. To do this navigate to Settings > Connections > Data usage > Data saver, then tap to turn Data saver on.

You can let some apps have unrestricted data usage by tapping Allow app while the Data saver feature is on. Then tap the Next button on each app to specify the restrictions.

Mobile hotspot

The Mobile hotspot makes use of your data plan to create a Wi-Fi network that is usable by different devices. Navigate to Settings > Connections > Mobile hotspot and tethering > Mobile hotspot. Then toggle on the Mobile hotspot. Activate Wi-Fi on the devices that you want to connect then choose the Mobile hotspot of your device. Finally, enter the password if required.

•Tap Connected devices to see a list of devices that are connected to your Mobile hotspot

Tethering

You can make use of tethering to share the internet connection of your device with other devices. Navigate from Settings > Connections > Mobile hotspot and tethering. Using a USB cable, connect the computer to your device and tap USB tethering

Nearby devices Scanning

Set up connections easily by activating the nearby device scanning feature. If there is an available connection nearby, this feature will send you a notification.
Navigate to Settings > Connections > More connection settings > Nearby device scanning then tap to turn it on.

Connect to a printer

Connect your device to a printer that is on the same Wi-Fi network to print images and documents from your device.

Navigate to Settings > Connections > More connection settings > Printing. Tap the Download plugin and follow the instructions to include a printing service. Finally, tap the printing service and More options > Add printer.

Be aware that not all apps support printing services.

Virtual Private Networks

With a Virtual Private Network (VPN), you can connect to a private and secured network with your device. To achieve this, you will need to obtain the connection information from your VPN administrator. Navigate to Settings > Connection > More connection settings > VPN > More option > Add VPN profile. Then input the information of the VPN as provided by the network administrator and tap Save.

Note that you will need a secured screen lock to configure your VPN.

VPN Management

Delete or modify the settings of your VPN using the VPN settings menu. Navigate to Settings> Connections > More connection settings > VPN. Then tap the Settings located next to a VPN, edit the VPN, and tap Save or remove the VPN by tapping Delete.

VPN Connection

Connecting to and disconnecting from a VPN is very easy as far as you have already set up the VPN. Navigate to Settings > Connections > More Connections Settings > VPN. Tap any VPN and input the information for login and tap Connect. To disconnect, simply tap the VPN and then tap Disconnect.

Ethernet

You can make use of an Ethernet cable to connect to a network if there is no available wireless network connection. To do this, first, connect an Ethernet cable to your device. Then navigate to Settings > Connections > More connection settings > Ethernet, then follow the instructions as given. Note that you will need an adapter to be able to connect an Ethernet cable to your device.

Vibration and Sounds

You can take control of the vibrations and sounds that are used to indicate screen interactions and notifications.

Sound mode

Without making use of the volume keys, you can still change the sound mode of your device.

Navigate to Settings > Sounds and Vibration and then select a mode:

•Sound: Make use of the vibrations, volume level, and sounds that you have chosen via the Sound settings for alerts and notifications.

–Vibrate while ringing: You can set your device to vibrate and ring at the same time when receiving a call.

•Vibrate: Make use of vibrations only for alerts and notifications.

•Mute: Set your device to be silent.

–Temporary mute: Place a time limit for the device to be muted.

Not to lose your customized sound levels, make use of the sound mode settings in place of the volume keys to modify the model of the sound.

Easy mute

Mute sounds fast by just turning the device over or covering the screen. Navigate to Settings > Advanced features > Motions and gestures > Easy mute and then tap to enable.

Vibrations

You can control the way and time that your device vibrates. To do this, navigate to Settings > Sounds and vibration > Options. The different options include:

•Vibration pattern: Choose from the already available vibration patterns.

•Vibration intensity: Drag the sliders to set the intensity of the vibration for notifications, calls, and interactions.

Volume

Set the level of the volume for notifications, ring tones, and other audio.

Navigate to Settings > Sounds and vibration > Volume, and then drag the sliders.

Note that can also adjust the volume using the Volume key. To customize all the options for volume, slide the volume controls.

Ringtone

Simply add your ringtone or choose from the preset ones to customize your ringtone. Navigate to Settings > Sounds and vibration > Ringtone. Tap any ringtone and listen to its preview then select your choice. Alternatively, tap Add to make use of an audio file as your ringtone.

System sound

Select a sound theme of choice for charging, Samsung Keyboard, and other touch interactions.

Navigate to Settings > Sounds and Vibration > System sound, then select any desired option.

System vibration and sounds

Customize the vibration and sound of your device for actions like charging and screen tapping.

Navigate to Settings > Sound and vibration > System sound/vibration control to view the following options:

System sounds

•Touch interactions: Anytime you touch the screen some tones will be Okayed to showcase selections.

•Screen lock/unlock: Anytime you lock or unlock the screen, a tone will be played.

•Charging: Anytime a charger is connected, a tone will be played.

•Dialing keypad: Anytime you dial numbers on the keyboard, a tone will be played.

•Samsung keyboard: Anytime you are typing on the Samsung keyboard, a tone will be played.

System vibration

•Touch interactions: Anytime you hold down buttons on the screen or tap the navigation buttons, your device vibrates.

•Samsung keyboard: When typing with the Samsung keyboard, your device vibrates.

Dolby Atmos

Relish the quality of the Dolby Atmos when you are playing content that has been mixed specifically for Atmos. This feature may be only available if the is a connected headset. To get the following options, navigate to Settings > Sounds and Vibration > Sound quality and effect:

•Dolby Atmos: Experience great audio that flows around and above you.

•Dolby Atmos for gaming: Make use of Dolby Atmos that is gaming-optimized.

Equalizer

Select an audio preset that is customized for different music genres. Alternatively, you can manually change the settings of your audio.

Navigate to Settings > Sounds and vibration > Sound quality and effect. The tap Equalizer to select a genre of music.

Audio options for Headset

Improve your music's sound resolution to get the best listening experience.

You can only access these features with a connected headset.

Navigate to Settings > Sounds and Vibration > Sound quality and effects, then turn it on by tapping any of the options:

•UHQ upscaler: This feature sharpens the audio resolution for a crisper sound.

Display

You can adjust the timeout delay, font size, the brightness of the screen, and other settings for the display.

Dark mode

The Dark mode feature lets you switch to a dark theme for a more comfortable night viewing. This darkens the bright or white screen and notifications.

Navigate to Settings > Display and choose any of the options below:

•Light: Select a light color theme for your device. This comes as the default option.

•Dark: Select a dark color theme for your device.

•Dark mode settings: Choose the time and when the Dark mode can be applied.

–Turn on as scheduled: Set the Dark mode for either Custom schedule or Sunset to Sunrise.

–Apply to wallpaper: This sets the Dark mode to the wallpaper anytime it is active.

–Adaptive color filter: Automatically switch on the Blue light filter between sunrise and sunset to reduce the strain on the eye.

Screen brightness

Modify the brightness of the screen according to your preference or lighting conditions.

Navigate to Settings > Display. To customize the brightness options:

•Set a custom brightness level by dragging the Brightness slider.

•Automatically adjust the brightness of the screen depending on the lighting conditions by tapping on Adaptive brightness.

Blue light filter

If you are one of those that make use of your device at night, the Blue light filter can help you to sleep better. You can automatically turn this feature off or on by using the Set a schedule option.

Navigate to Settings > Display > Blue light filter and select any of the options below:

•Set how opaque the filter will be by dragging the Opacity slider.

•To enable this feature, tap Turn on now.

•To set a schedule for the time that the blue light filter should be activated, tap Turn on as scheduled. Note that you can also select Custom schedule or Sunset to Sunrise.

Screen mode

They are different options for the screen mode available on your device. This adjusts the quality of the screen for different conditions. Choose any of the modes according to how you want it.

Navigate to Settings > Display > Screen mode. Then tap any of the options to set a screen mode.

Screen zoom

Modify the zoom level to decrease or increase the screen's content size.

Navigate to Settings > Display > Screen zoom. Then adjust the zoom level by dragging its slider.

Screen resolution

Lowering the resolution of the screen saves battery power while increasing it sharpens the quality of images.

Navigate to Settings > Display > Screen Resolution. Select the resolution of your choice and tap Apply.

Note that not all app support lower or higher screen resolutions. This may cause them to close if you change their designated resolution.

Full-screen apps

You can select the apps that you want to use in the full-screen aspect ratio.

Navigate to Settings > Display > Full-screen app, then tap apps to activate this feature.

Screen timeout

You can configure the screen to go off amount sometime.

Navigate to Settings > Display > Screen timeout, the tap a limit.

Note that, except in Always-On display, a prolonged display of static images can cause permanently degraded image quality. When you are not making use of the device, turn it off.

Accidental touch protection

Stop the device from registering touch inputs when it is inside a bag or pocket.

Navigate to Settings > Display> Accidental touch protection, then disable or enable it.

Touch sensitivity

This feature increases the sensitivity to touch of the device when being used with a screen protector.

Navigate to Settings > Display > Touch sensitivity, then enable.

Show information for charging

When the screen is on, the device can still display the battery level and estimated charging time.

Navigate to Settings > Display > Show charging information, then tap to enable.

Screen saver

While charging the device, you can also display photos and colors.

Navigate to Settings > Display > Screen saver and select any of the options below:

•None: Disable the screen saver.

•Colors: Tap the selector to show the screen displaying **changing colors**.

•Photo table: Display pictures in the format of a photo table.

•Photo frame: Display pictures in the format of a photo frame.

•Photos: Display pictures located in your Google Photos account.

3. To see a demonstration of the screen saver that you have selected, tap Preview.

To see more options, tap Settings located next to any feature.

Device maintenance

See the status of the storage, memory, and battery of your device. You can also optimize the system resources of your device.

Quick optimization

This feature improves the performance of your device through the following ways:

•It locates apps that make use of excess battery power and also clears unnecessary memory items.

•It deletes files that are not needed and also close running background apps.

•Malware scanning.

To make use of the quick optimization feature navigate to Settings > Device care > Optimize now.

Battery

See how different activities make use of your device's battery power. Navigate to Settings > Device Care > Battery to see the following options:

•Battery usage: See the usage of battery power by services and apps.

•Power mode: Choose a power mode to extend the battery life.

•App power management: Setup the way apps that are not frequently used utilize battery power battery.

•Wireless PowerShare: Set up wireless charging of your device battery with supported devices.

•Charging: To support fast charging, activate the following options:

–Fast charging

–Fast wireless charging

–Superfast charging

Storage

This displays the storage space and usage of the device.

You can unmount, mount or format a memory card.

Memory card

The memory of an installed micro SD memory card is display in the Storage setting. See Set up your device to know more.

Memory card mounting

A memory card, when installed, is automatically connected to the device for use. If for any reason you unmount the memory cards you have to mount them again before you can make use of them.

Navigate to Settings > Device Care > Storage > Advanced > Portable storage > SD card, then tap Mount.

Memory

Ascertain the available memory. To speed up the device, you can close background apps.

Navigate to Settings > Device care > Memory. The used and available memory space is display there.

• To free up memory space tap Clean.

• To see the list of apps making use of memory, tap View more. Then tap any app to include or exclude their services.

•Tap Apps not used recently to see the services and apps that are not included here then tap any of the apps to include or exclude them.

•To select the apps to exclude from using memory, tap Select apps to exclude.

Input and Language

Set up the input and language settings of your device.

Change the language of your device

You can include any language in your list and arrange them according to how you want it. If an app does not support your most preferred language, then it automatically moves to the next language in your list that is supported.

Navigate to Settings > General management > Language and input > Language. Tap add and choose any language from the list displayed. Then tap Set as default and finally Apply.

Time and Date

By default, your device receives information concerning time and fat from your wireless network carrier. You can manually set the time and date outside the network coverage.

Navigate to Settings > General management > Date and Time. Available there are the following options:

•Automatic date and time: Get the updates concerning time and date from your wireless network. If you disable the Automatic date and time feature, the following options will be available:

–Set date: Input the current date.

–Select time zone: Select a new time zone.

–Set time: Input the current time.

•Use 24-hour format: Set the time display format.

Troubleshooting

You can search for software updates and also reset some services as required on your device.

Software update

Search for and install any available device software update. Navigate to Settings > Software update to get the following options:

•Download and install: Check and install any available software updates.

•Last update: See the information concerning the current software's installation.

Reset

Reset your network and device settings. You can also factory reset your device.

Reset settings

A Reset resets everything to its default state, except the account, security, and language settings. Also, your data will not be affected.

Navigate to Settings > General management > Reset > Reset settings. Tap reset settings and follow the instructions.

Reset network settings

With Reset network settings, you can reset your Bluetooth, mobile data, and Wi-Fi settings.

Navigate to Settings > General management > Reset > Rest network settings, then tap Reset settings and follow the instructions displayed.

Factory Data Reset

Resetting your device to its factory default state erases every data from the device. It permanently erases application and system settings and data, music, photos, videos, accounts, and other files. The files and data stored on the external SD card will more be affected. The Factory Reset Protection (FRP) will be activated anytime you sign into a Google account on the device. This protects your device in case it is misplaced or stolen. To reset your device to factory default settings with an activated FRP, you have to input the username and password of the registered Google Account to gain access to the device. Without the correct credentials, you will not be able to access the device.

Note that after resetting the password of your Google Account, it may take a full day for it to sync with the devices that are registered to that account.

Security and Lock screen

Use a screen lock to secure and protect your device and data.

Screen lock types

Choose from the following security options: None, Password, PIN, Pattern, Swipe.

Note that there is also a biometric lock to protect sensitive data on your device. To learn more, see Biometric security.

Set a secure screen lock

It is ideal to secure your device with a secure lock (PIN, Pattern, or Password).

To set this up, navigate to Settings > Lock screen. Choose any screen lock type (Pin, Password, or Pattern) then tap on any to display the notifications.

The available options include:

View style: Either show the details of the notification Display hide them and only display an icon.

•Hide content: Do not display notifications in the Notification panel.

•Notifications to show: Select which notifications should be displayed on the Lock screen.

•Show on Always On Display: Show notifications on the Always-on Display screen. When you are finished, tap Done.

Google Play Protect

You can set up Google Play to periodically check your device and apps for any security threats.

Navigate to Settings > Biometrics and security > Google Play Protect.

•The updates are automatically checked for.

Find My Mobile

You can protect your device against theft or displacement by locking and tracking it online. This feature also enables you to delete data remotely. To make use of Find My Mobile, a Samsung account is needed. Also, the Google location service has to be enabled. To know more, visit samsung.com/us/support/owners/app/find-my-mobile

.

Turn on Find My Mobile

Turn on and customize the Find My Mobile feature before using it. Visit findmymobile.samsung.com to remotely access your device.

Navigate to Settings > Biometrics and security > Find My Mobile. Tap to turn it on, then log into your Samsung account. When this is done, you will find the following options:

•Remote unlock: Allow Samsung to save your pattern, password or PIN, to allow you to remotely control and unlock your device.

•Send last location: Enabling this allows your device to send its last location to the Find My Mobile server when the remaining battery level reaches a certain limit.

Set up SIM card lock

You can lock your SIM card with a PIN. This prevents unauthorized access to your SIM card.

Navigate to Settings > Biometrics and security > Other security settings > Set up SIM card lock. Then follow the instructions.

•Turn on this feature by tapping the Lock SIM card.

•To create a new pin, tap Change SIM card.

View passwords

You can enable characters to be briefly shown in the password fields as you input them.

To turn on this feature, navigate to Settings > Biometrics and security > Other security settings > Make password visible.

Device administration

You can set security authorizations to the different apps and services such as having administrative access to your device.

Navigate to Settings > Biometrics and security > Other security settings > Device admin apps. Then tap any option to switch it on as a device administrator.

Accounts

You can manage and connect to your accounts. These accounts include Samsung account, Google account, email accounts, and social media accounts.

Samsung Cloud

Keep your data safe by backing it up in the cloud when restoring your device. It is also possible to sync your data across different devices. Visit samsung.com/us/support/owners/app/samsung-cloud to learn more.

Navigate to Settings > Accounts and backup > Samsung Cloud. There will be prompts explaining to create or sign in to an account if there is no added account. You can then view and manage the items stored in the Samsung Cloud once an account has been set up.

Add an account

You can also sync your picture, video sharing, email, and social media accounts.

Navigate to Settings > Accounts and backup > Add account. Tap any of the accounts and follow the instructions to input your details and set up the account. You can auto-sync the day to activate auto-updates across all your accounts

Account settings

There are custom settings in each account. You can still set up a common setting across the accounts, though the account settings and features vary across the accounts.

Navigate to Settings > Accounts and backup > Accounts. Tap on any account to set up the settings for its synchronization. You can tap other available options.

Backup and restore

You can set up your device to back up data to your accounts.

Samsung account

You can activate information backup to your Samsung account.

To view the available options, navigate to Settings > Accounts and backup > Backup and restore:

•Back up data: Setup your Samsung account to back up your data.

•Restore data: Restore your backed up data using your Samsung account.

Google Account

You can activate information backup to your Google Account.

To view the available options, navigate to Settings > Accounts and backup > Backup and restore:

•Back up my data: Enable the back up of Wi-Fi settings and passwords, app data, and other data settings to Google servers.

•Backup account: Choose the Google Account that you want to be used as your back up account.

•Automatic restore: This enables settings to be automatically restored from Google servers.

Colors and clarity

For easier viewing, you can adjust the contrast and color of screen elements and text.

Navigate to Settings > Accessibility > Visibility enhancement and tap on any of the following options:

•High contrast theme: To enable easier viewing and increase the contrast, adjust the fonts and screen colors.

•High contrast fonts: To increase the contrast with the background, adjust the outline and color of the fonts.

•High contrast keyboard: To increase the contrast between the keys and the background, adjust the Samsung keyboard's size and also change its color.

•Show button shapes: Display the buttons with a shaded background. This makes them stand out against the wallpaper.

•Color inversion: Change the color display from black text on a white background to white text on a background and vice versa.

•Color adjustment: If see screen color is difficult, adjust the screen's color.

•Color lens: If you have difficulty in reading the text displayed, adjust the screen colors.

•Remove animations: If you are animation sensitive, do away with the screen effects.

Now, you can go all along to having an unending exciting moment with your extraordinary Samsung galaxy s20 ultra 5G device.